2nd Grade Science Volume 1

© 2013 Todd Deluca
OnBoard Academics, Inc
Newburyport, MA 01950

800-596-3175
www.onboardacademics.com

Table of Contents

The Changing Earth

Label the continents.

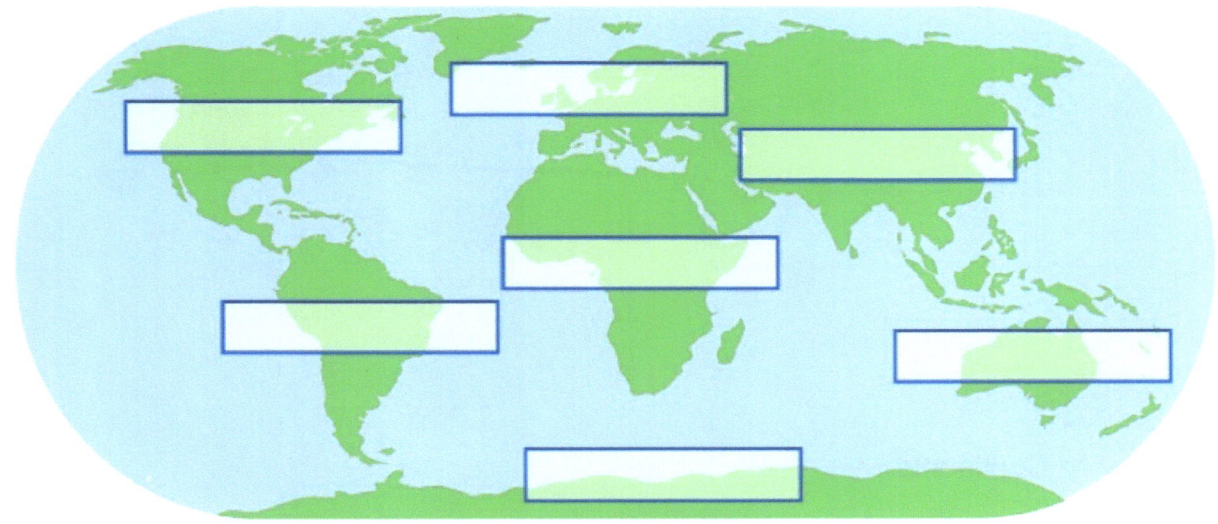

EUROPE　　　**AUSTRALIA**　　　**AFRICA**　　**ASIA**

NORTH AMERICA　　　**SOUTH AMERICA**　　　**ANTARCTICA**

Have you ever noticed that the continents of the earth look a bit like a jigsaw puzzle?

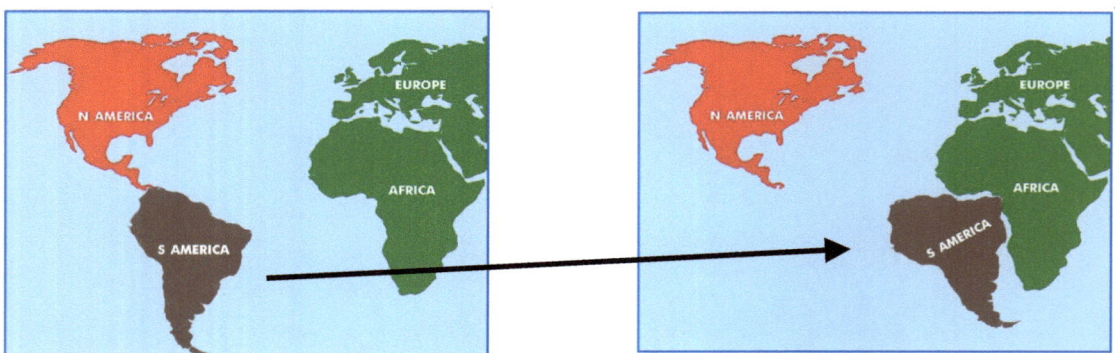

The Moving Continents

The German scientist, Alfred Wegener, suggested that 250 million years ago the Earth's continents were joined together in one supercontinent which was named Pangaea. There is evidence for Pangaea in both the fossil records and in the shape of the continents that we know today.

The Moving Continents

Scientists believe that the Earth's continents were once a single continent called Pangaea but have moved apart over millions of years. We call this continental drift.

This happens because the outer layer of the Earth is actually made of slabs of rock called tectonic plates that float on hot liquid rock called the mantel. As the tectonic plates drift, the continents that are attached also move. The continents are moving or drifting a couple centimeters a year even today.

This also explains earthquakes and volcanoes. Volcanoes occur when hot magma escapes through cracks in the earths surface caused by the shift in tectonic plates. When the magma erupts through the earths surface its called lava. The lava cools to form new rock. When a very large volcano erupts deep in the ocean new islands can form. Hawaii and Japan were both formed from volcanoes.

As the Earth's tectonic plates move, they sometime slide past each other. One may move under another or they may collide. This can cause earthquakes to occur. When two plates collide they may be forced together to form new mountains.

When they plates shift and collide they cause the Earth to shake and this can destroy large buildings.

Water and wind change the Earth's surface.

Wind and water are continually changing the Earth's surface. For example when it rains and water enters a crack in a rock and then freezes, the ice expands and breaks off parts of the rock. This is called weathering.

Running water and rain wear away and smooth rock. This process is called erosion. The Grand Canyon formed this way. The Colorado River gradually eroded away the rock and cut the canyon deeper and deeper.

Glaciers, which are mountains of moving ice with rock trapped within the ice also cause erosion by moving through a valley and widening it and smoothing it. You can see this in Yosemite National Park.

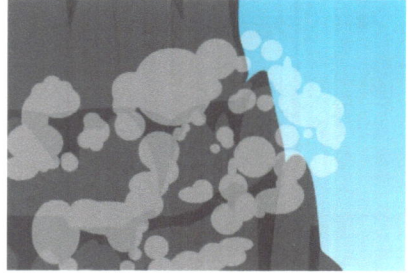

Wind also causes erosion and weathering by moving and depositing material to new places. Wind carries small bits of rock that are blown against other parts of the Earth's surface carving away and smoothing the surface.

Living things change the Earth's surface.

Living things can effect weathering and erosion. For example roots in a plant hold the soil in place and can prevent the rain and wind from washing away the soil.

On the other hand the roots can cause weathering ,the roots grow into rock's cracks causing them to crumble.

Earth worms dig holes that help the soil to get water. Other animals can cause problems by eating all the vegetation that enables easy erosion from water and wind.

People also have a big impact on erosion and weathering by building roads, cutting down trees and even recreational activities like hiking.

Humans also have a big impact on preventing erosion by planting trees, building break walls near the ocean and terracing hills to prevent run off.

Do I prevent weathering and erosion?

√ if I prevent weathering and erosion and X if I don't.

	I am going to climb this hill.
	I am a river running through a valley.
	I am going to plant some trees on a sloping hill.
	We built a breaker on the beach.
	I'm a large glacier.
	We keep to the trails when we go hiking.

The Changing Earth Quiz

1. The Earth's continents were once a single large continent called _____.
    a. Asia
    b. Pangaea
    c. Africa
    e. Europe

2. _____ is the process of continents moving apart over a period of time.
    a. tectonic plates
    b. earthquakes
    c. continental drift
    d. volcano

3. The continents are still moving today. True or false?

4. The surface of the Earth is made of several different slabs of rock called the mantle. True or false?

5. Erosion causes deposition of loose materials to new places. True or false?

Landforms

Which of these is a landform?

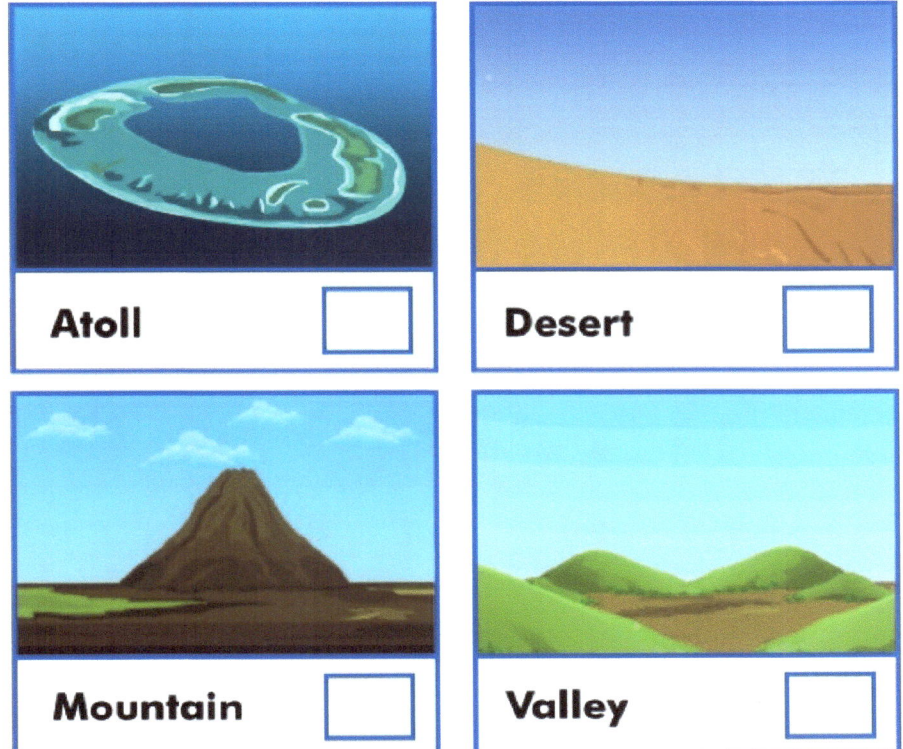

Atoll ☐   **Desert** ☐

Mountain ☐   **Valley** ☐

They all are! A **landform** is a type of land. There are lots of different landforms, some of which you probably have heard of such as a mountain or a waterfall; others which may be new to you, such as a plateau or a peninsula.

Landforms

Island
An island is an area of land that is surrounded by water

Atoll
An atoll is an island that is made form a ring of coral. In the middle of an atoll is a saltwater lake called a lagoon.

Bay
A bay is a part of the ocean that juts into the land.

Peninsula
A peninsula is like an island but is connected to the mainland by a thin strip of land.

Can you match each landform with its image?

atoll

bay

canyon

cliff

desert

glacier

marsh

oasis

swamp

cape

mesa

mountain

plain

plateau

peninsula

valley

volcano

waterfall

Sort these land forms by their elevation.

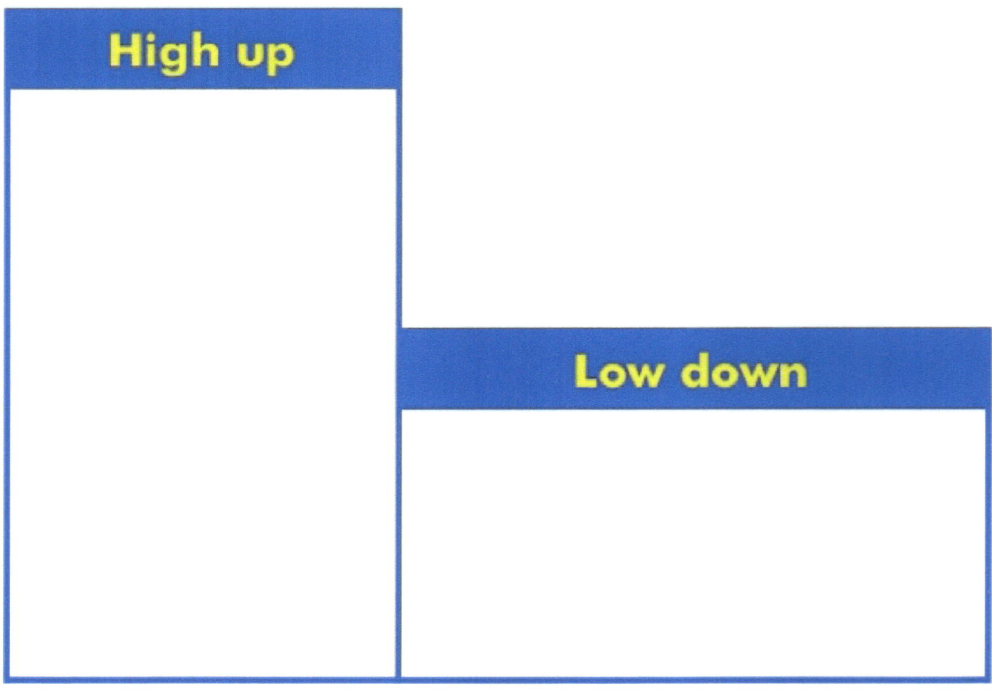

High up

Low down

canyon

plain

cliff

plateau

marsh

swamp

mesa

valley

mountain

Sort these landforms as always near the water or not near the water.

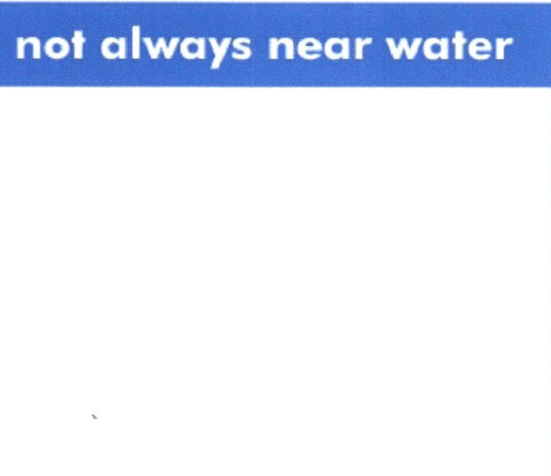

waterfall   peninsula   mountain   atoll   volcano

plain   swamp   valley   bay   oasis

mesa   marsh   cape   plateau

Famous North American Landforms

1 **Alaska Range**

2 **Rocky Mountains**

3 **Cascades**

4 **Sierra Nevada**

5 **Great Plains**

6 **Sierra Madres**

7 **Coastal Plain**

8 **Appalachian Mountains**

9 **Greenland Glacier**

Can you identify these famous mountain and deserts around the world?
Use the names below to label.

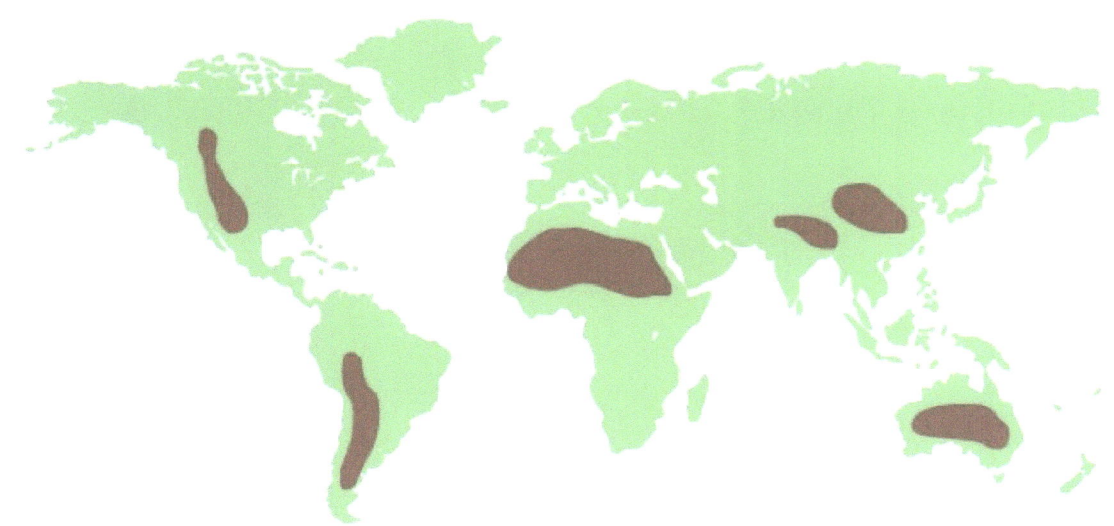

Himalayas   **Sahara Desert**   **Australian Deserts**

Andes   **Gobi Desert**   **Rocky Mountains**

Landforms Quiz

1. Deserts, mountains and valleys are types of landforms. True or false?

2. A _____ is a type of landform that is small than a plateau.
   cliff
   mesa
   cape

3. A _____ is a grassy land covered with a thin layer of water.
   bay
   glacier
   marsh

4. Which of these is an area of land completely surrounded by water on all sides?
   hill
   oasis
   island

The Water Cycle

Where has this water been before you drink it?

Underground   **In a glacier**   **In a lake**   **In a plant**

In a dinosaur   **In the ocean**   **In a river**   **In a cloud**

_____

Every glass of water you have ever drunk might have been in all of these places at one point in time. This is because all of the water on Earth is about as old as the Earth (over 4 billion years old) and is always moving between different places on Earth due to the water cycle.

How does water cycle from place to place?

There are essentially three ways that water moves from place to place; evaporation, condensation and precipitation. This might sound like a very complicated process but the water cycle is quite straight forward.

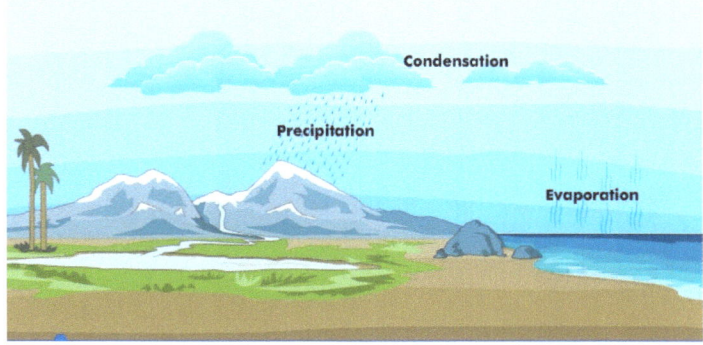

Lets start by looking at one drop of water in the ocean. Oceans are a great place to start because almost all of Earth's water (97%) is found in the ocean. An average drop of water spends 3,000 years in the ocean. But, if our drop of water gets close to the surface on a warm day it will get enough energy from the sun to evaporate. Evaporation is when water turns from a liquid into a gas that we call water vapor. Evaporation is the reason that a puddle dries during a warm day.

OK, so our water drop has now left the ocean and is floating in the air as water vapor. Water vapor is lighter than air so it rises. As it rises it will cool off

due to wind and cooler temperatures in the higher atmosphere, When this happens it will lose its heat energy form the sun and turn back into a water drop. This is caused condensation. So to summarize condensation is when water turns from a water vapor into liquid.

Our water droplet isn't alone. Our water droplet is hanging out with all of his buddies and they got together to form a cloud. At some point as they cool and there are too many of them the cloud can't hold them and the water in the air falls to the ground. This is called precipitation. This is when water droplets fall from the clouds to the ground.

Where will our water droplet end up.
It might go to an ocean, a lake, a river, the soil or the open mouth of a dog. We aren't sure where it will end up but we are sure that it won't stay there forever because the water cycle repeats endlessly.

_____

Label the water cycle. There is a new term that describes the water's movement on the ground after it precipitates.

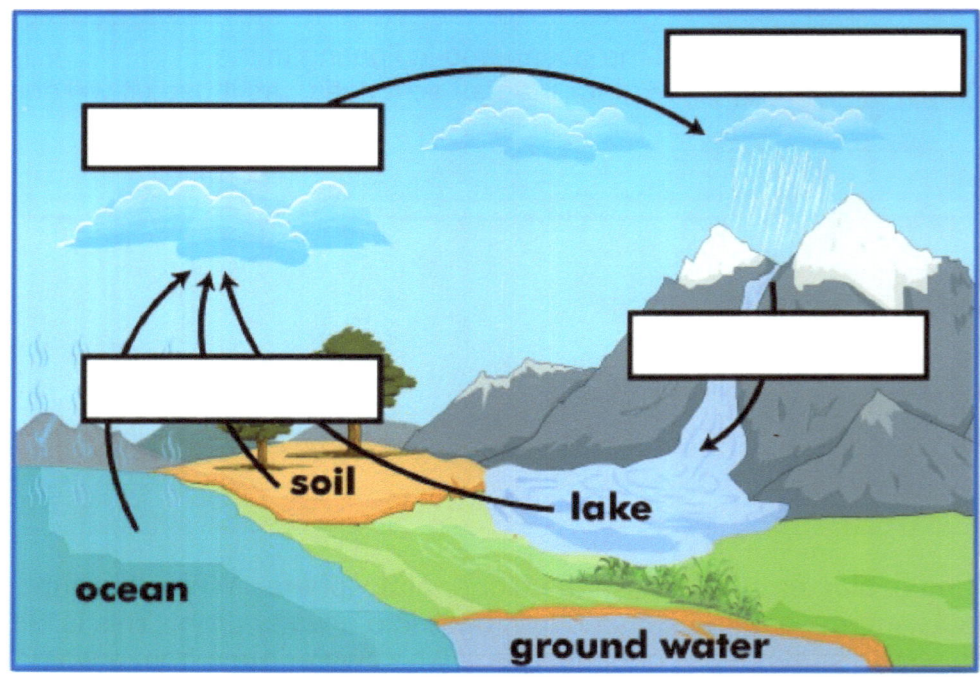

precipitation   condensation   evaporation   runoff

What is the state of water at each stage of the water cycle?

solid

liquid

gas

All plants and animals (including humans) need water, but all of the water we absorb or drink has already been used many times before. This includes water in juice, soda, and milk. The water that plants transpire or which you sweat or urinate will be re-used and re-drunk again many times over. The good news is that the water cycle cleans water, but pollution can sometimes disrupt the water cycle and the cleaning process.

Plants get water from the soil through their roots. They then release water in the form of water vapor through their leaves. This is a special type of evaporation called transpiration.

Animals get water from oceans, rivers and lakes using their mouths. They release the water through sweating or through urination. You might like to know that when urine evaporates, pure water evaporates into the air while the other chemicals stay behind or evaporate separately from the pure water.

Match the stage of the water cycle with the proper description.

Heat from the sun turns water close to the surface of oceans, lakes and rivers into a gas called water vapor.

condensation

Water vapor rises and cools and turns into liquid water.

evaporation

transpiration

Rain, snow sleet or hail fall from the clouds.

precipitation

Plants release water vapor through their leaves

The Water Cycle and Your Stove

From the colored suggestions write in what's happening in the illustration in the yellow box, the thermometer direction will give you a clue. In the white box, write in the chemical reaction from the clues on the right

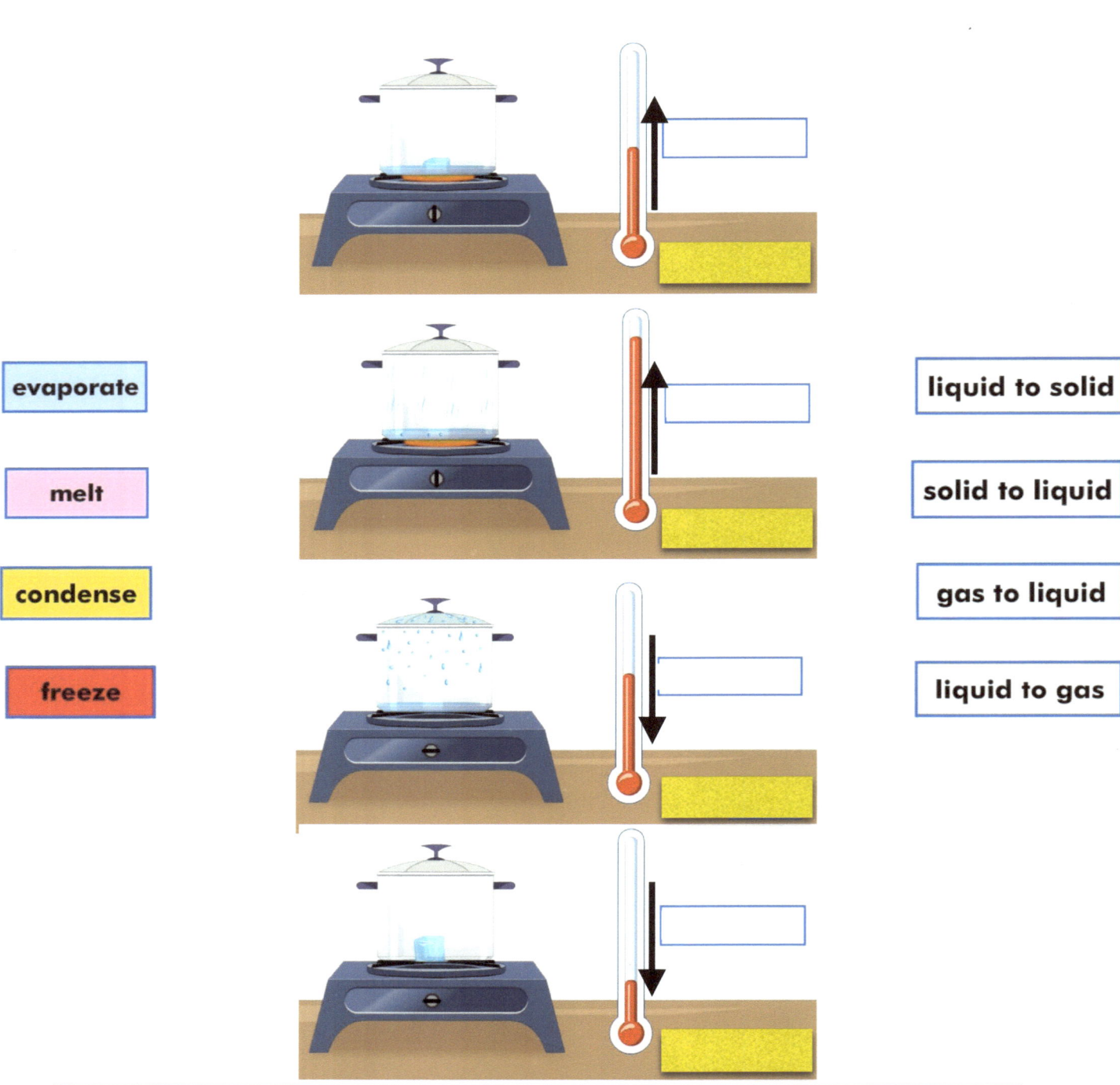

evaporate

melt

condense

freeze

liquid to solid

solid to liquid

gas to liquid

liquid to gas

A typical drop of water would spend approximately this amount of time at each location:
Animal or Plant: one day
Air: nine days
Soil: one month
River: four months
Lake: 75 years
Ocean: 3,200 years
Deep Underground: 10,000 years
Antarctica: 20,000 years

Evaporation or Condensation

Label each chemical reaction.

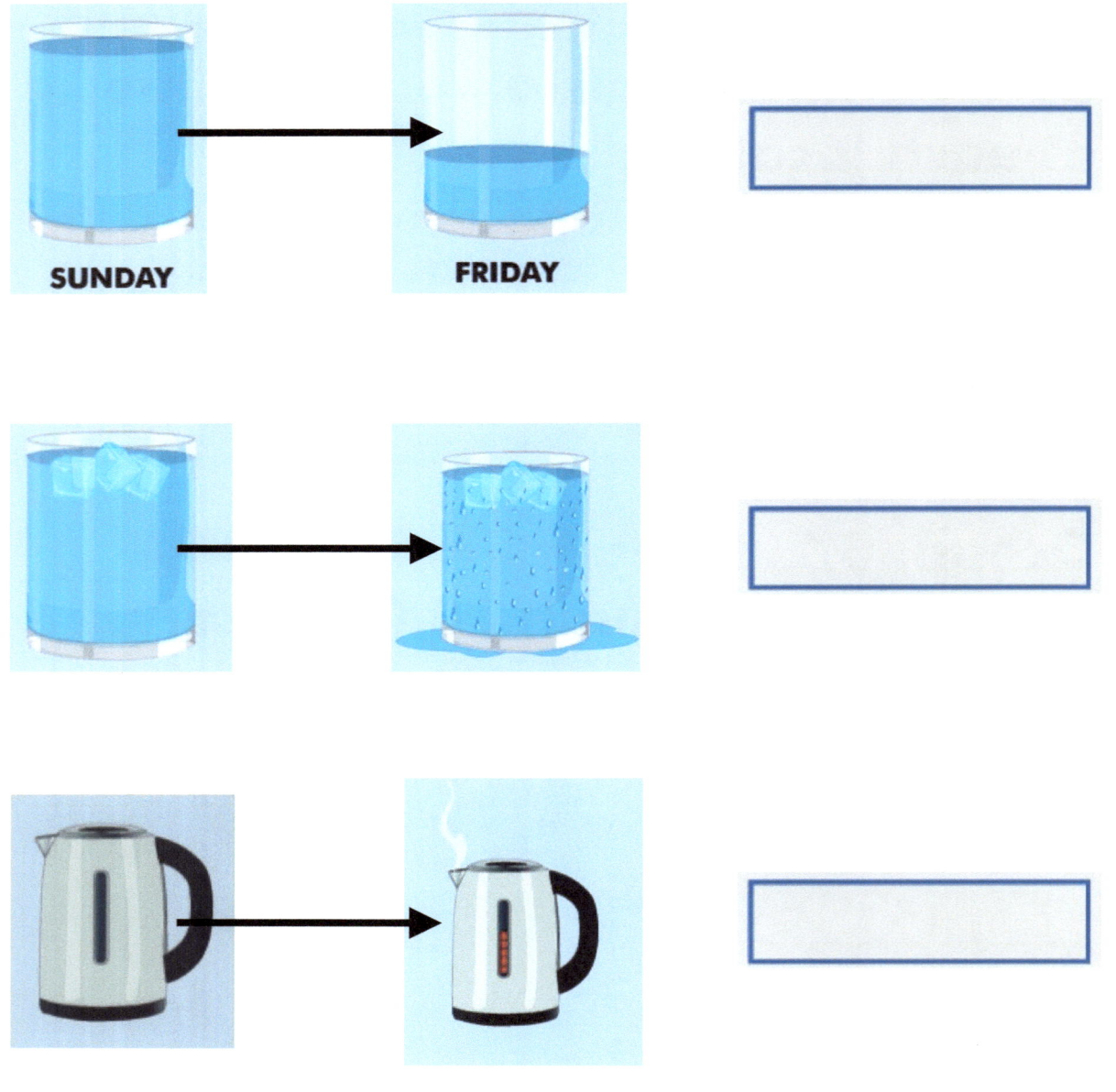

Water Cycle Quiz

1. Clouds are formed by _____.
    a. rising air
    b. condensing water vapor
    c. snow
    d. dew

2. The continuous movement of water on, above and below the surface of the Earth is called the water cycle. True or false?

3. When a cloud becomes too heavy , the water droplets come down as _____.
    a. vapor
    b. rain
    c. wind
    d. clouds

4. Evaporation primarily takes place during the day. True or false?

5. The process by which water changes from a liquid state to a gaseous state is called _____.

Clouds

How do clouds form?

Clouds are formed when energy from the sun heats water that is close to the surface of lakes, rivers and oceans and turns the liquid water into an invisible gas called water vapor. We call this process evaporation. Because water vapor is lighter than air it rises. As it rises it cools and turns back into water droplets.

We call this process condensation. If the air temperature below freezing or 00 c.
Then the water vapor condenses into tiny ice crystals. Clouds form when billions of these tiny ice crystals form around tiny dust particles in the air. Although heavier than air, they are so tiny and light that slight movement in

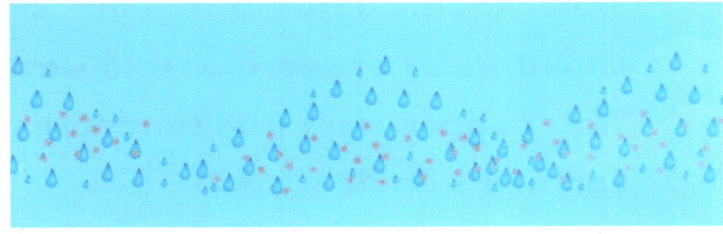

the air keep them floating. However when the water droplets and ice crystals get too heavy we get precipitation. This is the name we give to rain and snow.

There are a number of different types of clouds but they all form basically the same way. When water vapor condenses into liquid water or ice around particles of dust.

Fog is the name we give to clouds that form close to the Earth's surface.

> **Clouds form when water vapor condenses into tiny water droplets or ice crystals around particles of dust in the air.**

Order the five steps of cloud formation.

Water droplets cluster together around particles of dust.

Liquid water evaporates into water vapor.

Water vapor cools and condenses into liquid water droplets.

Sun heats liquid water.

Water vapor rises high into the atmosphere.

Three main types of clouds

> Cloud names are combined when a cloud has the characteristics of more than one cloud. For example, cirrostratus clouds are like layers of cirrus clouds. Similarly, stratocumulus clouds are cumulus clouds that cover the whole sky. Cirrocumulus clouds are small cumulus clouds that are high in the sky and contain ice.

cirrus

Cirrus means curl and cirrus clouds are thin curly feather like clouds. They are the highest clouds in the sky from about 6-13 kilometers high. They are so high that they contain ice crystals.

stratus

Stratus means layer and stratus clouds are the gray clouds that often blanket the entire sky. Stratus clouds are the lowest clouds in the sky usually below 2 kilometers. They have indistinct edges.

cumulus

Cumulus means heap. Cumulus clouds are the thick puffy clouds that look like cotton balls heaped on one another. These clouds also occur at or below about 2 kilometers.

Label these clouds.

| cumulus | stratus | cirrus |

Clouds Quiz

1. There are three main types of clouds in the sky. True or false?

2. _____ clouds float at 18,000 ft in the sky.

3. Clouds are made of water droplets or ice crystals. True or false?

4. Cirrus clouds are usually made up of ice crystals. True or false?

5. Cumulus clouds are dark clouds that predict storms and heavy rain/snow. True or false?

6. When the weather is foggy, it could be a really low _____ cloud.
   a. stratus
   b. cumulus
   c. cirrus
   d. cumulonimbus

Newburyport, MA 01950

1-800-596-3175

OnBoard Academics employs teachers to make lessons for teachers! We create and publish a wide range of aligned lessons in math, science and ELA for use on most EdTech devices including whiteboard, tablets, computers and pdfs for printing.

All of our lessons are aligned to the common core, the Next Generation Science Standards and all state standards.

If you like our products please visit our website for information on individual lessons, teachers licenses, building licenses, district licenses and subscriptions.

Thank you for using OnBoard Academic products.

www.onboardacademics.com